GOODS OF THE MIND, LLC

Competitive Mathematics Series

for

Gifted Students in Grades 1 and 2

PRACTICE OBSERVATION AND LOGIC

Cleo Borac, M. Sc.
Silviu Borac, Ph. D.

This edition published in 2014 in the United States of America.

Editing and proofreading: David Borac, M.Mus.
Technical support: Andrei T. Borac, B.A., PBK

All rights reserved. Except as permitted under the United States Copyright Act, no part of this publication may be reproduced or distributed in any form or by any means, or stored in a database or retrieval system, without prior written permission from the publisher, unless otherwise indicated.

Copyright Goods of the Mind, LLC.

Send all inquiries to:

Goods of the Mind, LLC
1138 Grand Teton Dr.
Pacifica
CA, 94044

Competitive Mathematics Series for Gifted Students
Level I (Grades 1 and 2)
Practice Observation and Logic
2nd edition

Contents

1 Foreword   5

2 Observation   8

3 Practice One   9

4 Reasoning   14

5 Practice Two   16

6 Rates   20

7 Practice Three   21

8 Logic   23

9 Practice Four   25

CONTENTS *CONTENTS*

10 Miscellaneous Practice 27

11 Solutions to Practice One 35

12 Solutions to Practice Two 42

13 Solutions to Practice Three 48

14 Solutions to Practice Four 52

15 Solutions to Miscellaneous Practice 56

Foreword

The goal of these booklets is to provide a problem solving training ground starting from the earliest years of a student's mathematical development.

In our experience, we have found that teaching how to solve problems should focus not only on finding correct answers but also on finding better solution strategies. While the correct answer to a problem can typically be obtained in several different ways, not all these ways are equally useful for learning how to solve problems.

The most basic strategy is *brute force*. For example, if a problem asks for the number of ways Lila and Dina can sit on a bench, it is easy to write down all the possibilities: Dina, Lila and Lila, Dina. We arrive at this solution by performing all the possible actions allowed by the problem, leaving nothing to the imagination. For this last reason, this approach is called brute force.

Obviously, if we had to figure out the number of ways 30 people could stand in a line, then brute force would not be as practical, as it would take a prohibitively long time to apply.

Using brute force to obtain the correct answer for a simpler problem is not necessarily a useful learning experience for solving a similar problem that is more complex. Moreover, solving problems in a quantitative manner, assuming that the student can transfer simple strategies to similar but more complex problems, is not an efficient way of learning problem solving.

From this simple example, we see that the goal of *practicing* problem solving is different from the goal of problem solving. While the goal of problem solving is to obtain a correct answer, the goal of practicing problem solving is to acquire the ability to develop strategies, generate ideas, and combine approaches that are powerful enough to solve the problem at hand as well as future similar problems.

While brute force is not a useless strategy, it is not a key that opens every door. Nevertheless, there are problems where brute force can be a useful tool. For instance, brute force can be used as a first step in solving a complex problem: a smaller scale example can be approached using brute force to help the problem solver understand the mechanics of the problem and generate ideas for solving the larger case.

All too often, we encounter students who can quickly solve simple problems by applying brute force and who become frustrated when the solving methods they have been employing successfully for years become inefficient once problems increase in complexity. Often, neither the student nor the parent has a clear understanding of why the student has stagnated at a certain level. When the only arrows in the quiver are guess-and-check and brute force, the ability to take down larger game is limited.

Our series of books aims to address this tendency to continue on the beaten path - which usually generates so much praise for the gifted student in the early years of schooling - by offering a challenging set of questions meant to build up an understanding of the problem solving process. Solving problems should never be easy! To be useful, to represent actual training, problem solving should be challenging. There should always be a sense of difficulty, otherwise there is no elation upon finding the solution.

Indeed, practicing problem solving is important and useful only as a means of learning how to develop better strategies. We must constantly learn and invent new strategies while questioning the limitations of the strategies we are using. Obtaining the correct answer is only the natural outcome of having applied a strategy that worked for a particular problem in the time available to solve it. Obtaining the wrong answer is not necessarily a bad outcome; it provides insight into the fallacies of the method used or into the errors of execution that may have occured. As long as students manifest an interest in figuring out strategies, the process of problem solving should be rewarding in itself.

Sitting and thinking in a focused manner is difficult to train, particularly since the modern lifestyle is not conducive to adopting open-ended activities. This is why we would like to encourage parents to pull back from a quantitative approach to mathematical education based on repetition, number of completed pages, and the number of correct answers. Instead, open up the

CHAPTER 1. FOREWORD

time boundaries that are dedicated to math, adopt math as a game played in the family, initiate a math dialogue, and let the student take his or her time to think up clever solutions.

Figuring out strategies is much more of a game than the mechanical repetition of stepwise problem solving recipes that textbooks so profusely provide, in order to "make math easy." Mathematics is not meant to be easy; it is meant to be interesting.

Solving a problem in different ways is a good way of comparing the merits of each method - another reason for not making the correct answer the primary goal of the activity. Which method is more labor intensive, takes more time or is more prone to execution errors? These are questions that must be part of the problem solving process.

In the end, it is not the quantity of problems solved, the level of theory absorbed, or the number of solutions offered in ready-made form by so many courses and camps, but the willingness to ask questions, understand and explore limitations, and derive new information from scratch, that are the cornerstones of a sound training for problem solvers.

These booklets are not a complete guide to the problem solving universe, but they are meant to help parents and educators work in the direction that, aside from being the most efficient, is the more interesting and rewarding one.

The series is designed for mathematically gifted students. Each book addresses an age range as some students will be ready for this content earlier, others later. If a topic seems too difficult, simply try it again in a couple of months.

OBSERVATION

One of the most important skills in problem solving, observation trains patience and lateral thinking.

Experiment

Which of the following figures on the right is not a larger version of the corresponding figure on the left?

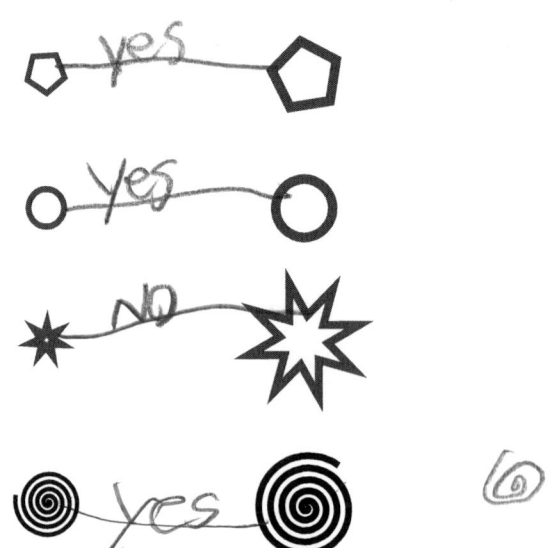

Even though the pentagon and the spiral are rotated, the larger figure is still only a bigger version of the smaller figure. The small star, however, has 7 corners while the large star has 8.

If the student rushes, he or she may indicate the spiral or the pentagon as dissimilar. If the student is patient and counts the spikes of the stars, he or she will get the correct answer.

Practice One

Exercise 1
Lila has a bracelet made of blue and white magnetic beads:

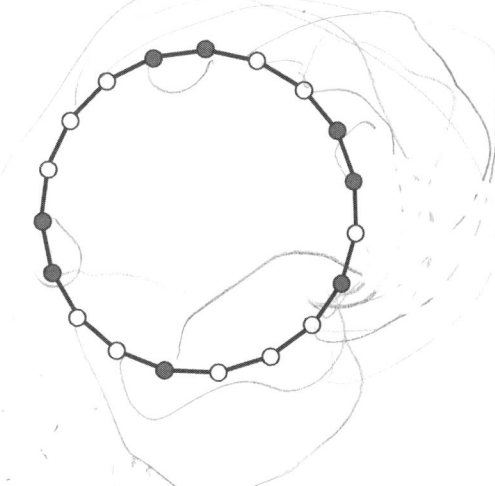

Lila would like to change the bracelet so that all beads of the same color are next to each other. She wants to do this by swapping two beads at a time. What is the smallest number of swaps she must make?

Exercise 2
Dina folds the following square along one of the grid lines:

She wants to obtain the largest number of matches for the small squares after folding. A match indicates that two squares of the same color overlap each other.

What is the largest number of matches she can get by cleverly choosing the line for the fold?

Exercise 3

Lila's dog lives in a rickety old doghouse. It has many holes in the roof!

Lila wants to use some old roofing patches to fix the doghouse. Each patch looks like this:

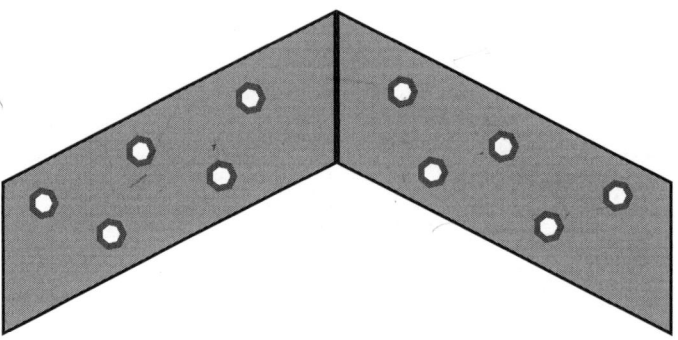

Lila cannot cut the patches but she can, if needed, overlap them. At least how many patches must she use in order to patch all the holes?

CHAPTER 3. PRACTICE ONE

Exercise 4

Which of the digits in the following picture is the largest?

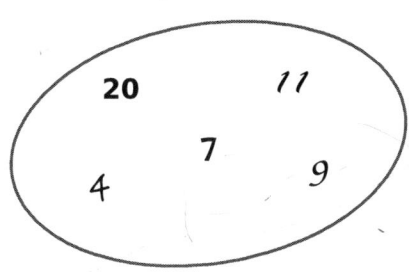

Exercise 5

Out of the pierced chips shown, Dina wants to build the highest tower that will have three holes that match on all chips. How tall will the tower be?

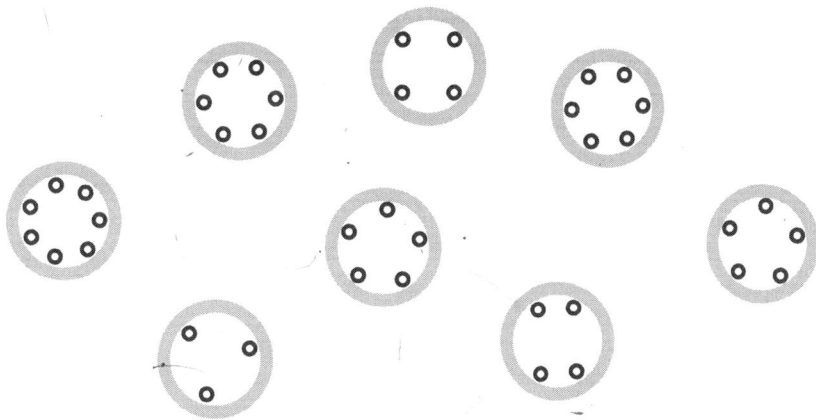

Exercise 6

How many even numbers can Lila make by removing some of the digits in the following string? Lila is not allowed to swap any of the remaining digits.

4 7 0 5

Exercise 7

There is a rule for forming the following set of numbers. In one of the numbers, some digits are missing. What could the missing digits be?

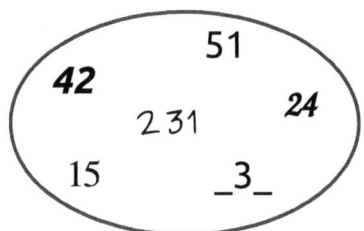

Exercise 8

Dina is allowed to swap any two adjacent (neighboring) digits in the following string. What is the smallest number of swaps she needs to make in order for all the "3"s to be one beside the other?

3 2 3 1 3 4 3

CHAPTER 3. PRACTICE ONE

Exercise 9

Amira is turning 4 years old. Her birthday cake has been sliced into quarters with one candle for each slice, like in the diagram:

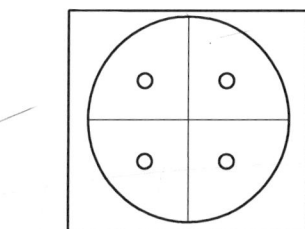

She asks her friend to move two candles at the same time and place each of them on a slice neighboring the slice it is on. What is the smallest number of such operations her friend must make in order to move all the candles to the same slice?

Exercise 10

Each of the figures in the diagram follows the same pattern. What number should be placed instead of the question mark?

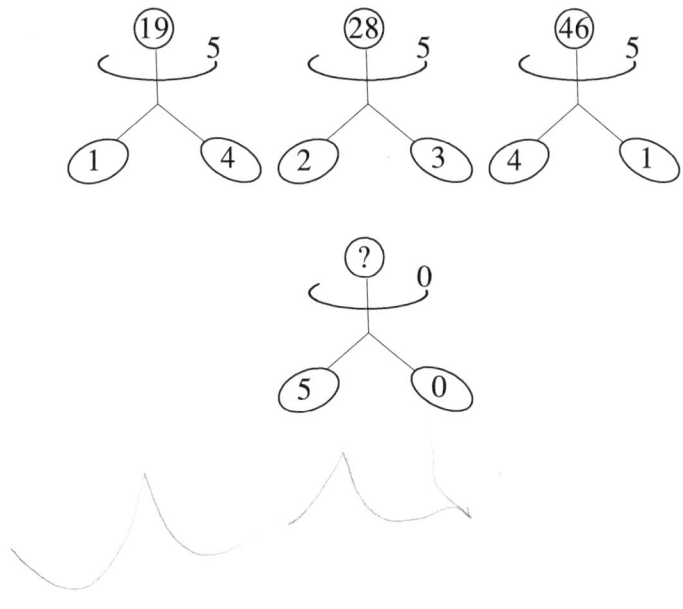

Reasoning

> **Note to parents:**
>
> Students are often trained to apply a certain operation to the numbers that come up in the problem. Many feel compelled to use simple arithmetic on the data given, but reasoning should be used to check the plausibility of the result. Questions to be asked include:
>
> 1. Can it be this large?
> 2. Can it be this small?
> 3. Is there some limitation to the process?
> 4. Are we answering the question asked or some other question?
>
> The underlying strategy for reasoning problems is to make sure the abstract model built using mathematics matches the physical situations it models.
>
> Reasoning requires attention to detail and good verbal or visual comprehension.
>
> More importantly, reasoning requires the ability to understand the specifications of the problem. Problems are extremely precise and are expressed in a special language that precludes misinterpretations. Students must start from the earliest grades to understand the language of problem statements and make it their own.

CHAPTER 4. REASONING

Example:

Answer the following questions:

1. Lila has 12 candles. She cuts each candle in half. How many candles does she have now?
2. Dina has 12 matches. She cuts each match in half. How many matches does she have now?
2. Amira has 12 envelopes. She cuts each envelope in half. How many envelopes does she have now?

Answers:

Notice that the physical nature of the objects we cut in half is important.

1. Lila will have 24 smaller candles.
2. Dina will have 12 shorter matches. Since the matches have only one end that produces a flame, each cut will produce a shorter match and a simple stick.
3. Amira will have 0 envelopes left. By cutting the envelope in two, the envelope is destroyed.

Practice Two

Exercise 1
Dina is making orangeade for 10 of her friends in a large pitcher with a capacity of 8 cups. She has 3 cups of cold water in the pitcher. To this, she adds 7 cups of orange juice. How many cups of orangeade does the pitcher contain now?

Exercise 2
Lila tied her dog Arbax to a small tree using a 6 foot line. At lunch time, Lila placed some bones 10 feet away from Arbax and went to picnic with Dina nearby. Can Arbax reach the bones?

Exercise 3
Dina and Lila are throwing a party for their birthday. They have a bag of identical party balloons to inflate. Using a small pump, Lila was able to inflate a balloon in 2 minutes. Using a larger pump, Dina was able to inflate a balloon in 1 minute. Whose balloon has more air inside, Dina's or Lila's?

Exercise 4
Amira is going down the stairs. Which of her feet is on the lower step: the right one or the left one?

CHAPTER 5. PRACTICE TWO

Exercise 5
Dina's phone shows the wrong time. It is 2 hours ahead of local time. If Dina's phone shows it is 11 pm, what is the actual time?

Exercise 6
Lila has three boxes and four potatoes. She places the potatoes in the boxes. Then, she makes up a number by counting the potatoes in each box. The leftmost box holds the number of hundreds, the middle box holds the number of tens, and the rightmost box holds the number of units, like in this example:

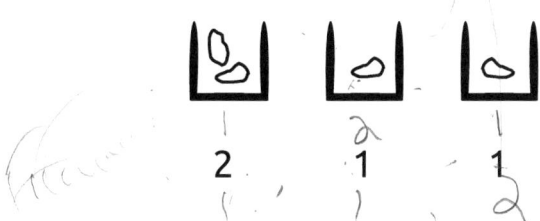

How many different numbers can she form?

Exercise 7
Lila is standing in line to buy popcorn at the cinema. There are 5 people in front of her and 3 people behind her. Each time someone

buys popcorn and leaves the line, two people join the line. How many people are standing in line by the time Lila gets to order?

Exercise 8

Dina has two cups of hot chocolate milk at 90 degrees Fahrenheit. What is the temperature of each cup?

Exercise 9

When the day before yesterday was tomorrow, it was Tuesday. Tomorrow it will be:

(A) Tuesday
(B) Wednesday
(C) Thursday
(D) Friday
(E) Saturday

Exercise 10

The purple warbler has a nest full of chicks. Half of her chicks have grey heads, half of her chicks have black tails, and half of her chicks have red bellies. The number of chicks cannot be:

(A) 2
(B) 4
(C) 5
(D) 6

Exercise 11

Arbax, the Dalmatian, is 10 steps away from Dina. Dina could be:

(A) anywhere on a 20 step long segment;
(B) anywhere on a 10 step long segment;
(C) at one of the ends of a 20 step long segment;
(D) anywhere on a square with a side 10 steps long;
(E) anywhere on a circle with a radius 10 steps long.

CHAPTER 5. PRACTICE TWO

Exercise 12

Lila and Dina follow the rules:

(i) replace two adjacent pluses with a minus

(ii) replace two adjacent minuses with a plus

(iii) replace two adjacent symbols that are different with a plus

to reduce the following string to a single symbol.

$$+ - + + -$$

Dina works from left to right. Lila works from right to left. Which one of them ends up with a plus: Dina, Lila, or both?

Rates

A *rate* is a quantity that describes how quickly another quantity changes over *time*.

In elementary problem solving we only handle *constant* rates. We are not concerned with rates that change. This means that if a problem states that a horse travels 20 miles in one hour, we assume that the horse starts directly at a speed of 20 mph and not from rest. This is, of course, not physically correct, but it is a very good model for handling average speeds.

At this level, students must be able to differentiate a situation in which a constant rate applies from a situation in which it does not.

For problems in which constant rates do apply, simple multiplications or divisions are needed. Therefore, the students should be at least in second grade.

Practice Three

Exercise 1
Dina runs 1 mile in 10 minutes. Lila runs half a mile in 6 minutes. Which of them runs faster?

Exercise 2
Arbax, the Dalmatian, barks 3 times every 10 minutes. How many times will Arbax bark in one hour?

Exercise 3
Each minute, Dina erases 4 letters from the left while Lila erases 3 letters from the right. After how many minutes will the board be wiped clean?

Exercise 4
It takes Dina 3 minutes to squeeze enough oranges to fill a glass with fresh juice. How many minutes will it take her to fill 5 glasses?

Exercise 5
It takes Lila 3 minutes to hardboil an egg. How long does it take her to hardboil 5 eggs?

Exercise 6
Two kittens can lap up a saucer of milk in 6 minutes. How long will it take four kittens to do the same?

CHAPTER 7. PRACTICE THREE

Exercise 7

One clock is fast and advances 10 minutes more than a regular clock every hour. At 4:00 pm, it is set to show the correct time. When the faster clock is one hour ahead of it, what time will the regular clock show?

Exercise 8

One person wears one pair of pants. How many pairs of pants will two pairs of people wear?

Exercise 9

It takes Dina three days to knit a scarf and it takes Lila two days to knit a scarf. How many days do the twins need to knit 5 scarves?

Exercise 10

Dina has a magical beanstalk that triples in height each day. In six days, the beanstalk reached the height of her house: 18 feet. How many days before that was the beanstalk only 6 feet tall?

Exercise 11

Over the summer, Lila stays with her cousins while Dina stays with her grandparents. They write each other a letter every day. How many letters will they send in a week?

Exercise 12

A car is 8 times faster than a bicycle. If it takes the car 1 hour to travel a circular path, how many laps around the path can the bicycle complete in 8 hours?

LOGIC

To solve a logic problem, a student may have to make a table to summarize information.

Example There are 3 marbles, one green, one orange, and one blue, in a box. One is made of plastic and two are made of glass. The plastic one is not green. The green and the blue marbles are made of the same material. Which one is made of plastic?

	green	orange	blue
plastic	no	?	?
glass	yes	?	yes

Notice how, after completing the second line, it becomes clear that the orange marble must be made of plastic.

The student must be able to recognize *exclusive* properties and *non-exclusive* properties.

Example John, Marie, and Maggie are friends. Some of them like apples and some of them like pears. Some of them have short hair and some of them have long hair.

The sum of the number of children who have long hair and the number of children who have short hair *must* equal 3, since no one can have both long and short hair at the same time.

The sum of the number of children who like apples and the number of children who like pears, however, may exceed 3 if there are any children who like both pears and apples.

Another solving strategy is to represent several categories of objects in a diagram.

Example There are 5 finalists at a dog show. Of the finalists, 4 have long hair and 3 are good runners. At least how many are good runners with long hair?

Make a diagram separating the dogs into two groups: long-haired and short-haired.

$$L\ L\quad S$$
$$L\quad\ \ L$$
$$L$$

In order to find the smallest number of long-haired dogs that are good runners, we mark the short-haired dog as a good runner first. Since there are more good runners than short-haired dogs, we now mark as many long-haired dogs as possible as good runners. Don't forget, however, that there are only 3 good runners among the 5 dogs. In the following diagram, the good runners have been circled:

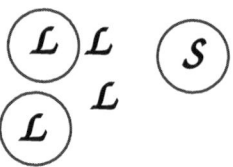

Thus the smallest possible number of good runners with long hair is 2.

Note: the diagram does not need to contain complicated drawings. Simple symbols will do the job.

Practice Four

Exercise 1

The friends of a friend are my friends. This is true for all the people in my circle of friends. If each of us has 8 friends, how many of us are there?

Exercise 2

Seven frogs are sitting on some waterlilies. Suddenly, a vulture swoops down and snatches one of the frogs. How many frogs are left sitting on the waterlilies?

Exercise 3

Dina and Lila are holding the baby while their father helps the nurse set up the weighing station. At least how many people are there in the room?

Exercise 4

Dina and Lila traveled by omnibus train from Blois to Tours in 4 hours. If they had used an express train instead, the trip would only have lasted 2 hours. Which train traveled a longer distance?

Exercise 5

Dina baked a large brownie cake and cut it in 12 slices. Lila cut one of the slices in 4. How many slices are there now?

Exercise 6

Lila wants to adopt a cat. At the animal shelter there are 31 cats, of which 15 are tabby and the rest are black. The cats are so cute that Dina also decides to adopt a cat. Dina and Lila walk out, each holding the cat of her choice. What is the largest possible number of black cats that remained at the shelter after Dina and Lila left?

Exercise 7

4 race cars covered 4 miles in 1 minute. How many seconds will it take a single race car to cover 4 miles?

Exercise 8

Dina has two friends in grade 2, five friends in grade 4, and four friends in grade 3. At least how many friends does Dina have?

Exercise 9

Four of Lila's friends like Hathi, the elephant. Four of Lila's friends like Rikki, the mongoose. Four of Lila's friends like Felix, the cat. At least how many friends does Lila have?

Exercise 10

Dina has four big magnets and four small magnets. Two magnets are red, one is blue, three are green, and two are yellow. No large magnets are red. No small magnets are blue. An equal number of small and large magnets are painted in the same color. How many large green magnets are there?

Exercise 11

Lila has 28 fish in her 30 liter fish tank. One day, she removes 15 liters of water from the tank. How many fish are there in the tank now?

Exercise 12

Dina has more books than Nina. Nina has fewer books than Lila. Edna has more books than Lila. Dina does not have the largest number of books. Who has the smallest number of books?

Miscellaneous Practice

Exercise 1
Dina is in Ms. Left's class at the left end of the schoolyard. Lila is in Ms. Right's class at the right end of the schoolyard. In the middle of the schoolyard, there is a flag. After school, Lila runs towards Dina as Dina walks slowly towards Lila. When they meet, which one of them is closer to the flag?

Exercise 2
Dina's father is a pilot. His plane can fly at 500 miles per hour. He takes off from Plampaloosa and flies for one hour. After this time, how far away from Plampaloosa can he *not* be?

(A) 0 miles

(B) 200 miles

(C) 300 miles

(D) 500 miles

(E) 600 miles

Exercise 3
In Crazy Horse Gulch, there is an unusual weather pattern. It rains for three days, is sunny for three days, rains again for three days, and so on. If it rained today and yesterday, what will the weather be like in 4 days?

(A) rainy for sure

(B) sunny for sure

(C) could be either rainy or sunny

Exercise 4
Lila's father is 30 years older than she is. How many years older than her will her father be when Lila is 30 years old?

Exercise 5

It takes a magical fig tree 8 years to grow 10 feet tall. How many years does it take 2 magical fig trees to grow 10 feet tall?

Exercise 6

Dina and Lila harvested a lot of figs this summer and decided to dry them out in the sun. If it takes 3 days to dry 100 figs, how many days does it take to dry 300 figs?

Exercise 7

Lila and Dina ran towards each other on a 6 mile road. Dina ran twice as fast as Lila. By the time they met, how many miles had Lila run?

Exercise 8

An animation begins with a row of 25 squares, all of which are white. Every second, some of the squares turn blue. First, every second square starting from 1 turns blue. Then, every third square starting from 1 turns blue. Then, every fourth square starting from 1 turns blue, and so on. When will the entire row be blue?

Exercise 9

A *hexagon* is a closed figure with 6 sides. How many hexagons are hidden in the picture?

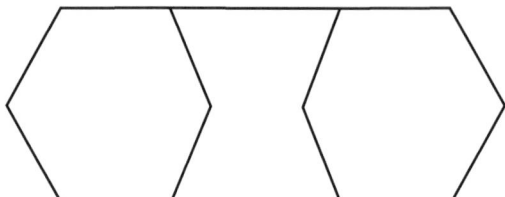

CHAPTER 10. MISCELLANEOUS PRACTICE

Exercise 10

Dina has lots of these tiles:

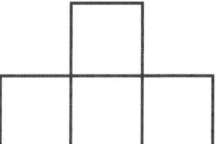

How many of the figures below can she make?

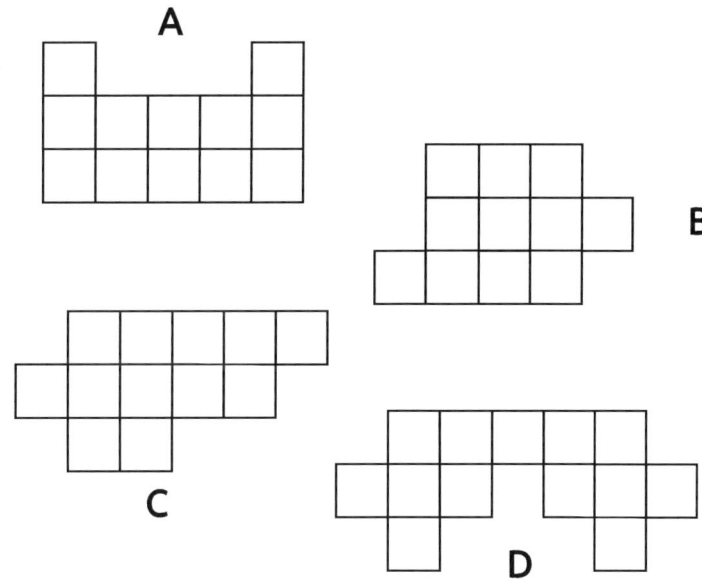

CHAPTER 10. MISCELLANEOUS PRACTICE

Exercise 11

Lila received a robomouse for her birthday. It was designed to walk on a grid either horizontally or vertically. Unfortunately, Lila's robomouse is defective and unable to turn left or go backwards. When Lila places it in the labyrinth at point A, facing in the direction of the arrow, it gets stuck at B.

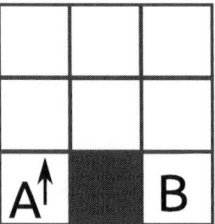

If Lila places it at point B facing in the direction of the arrow, at which of the following points will it get stuck?

CHAPTER 10. MISCELLANEOUS PRACTICE

Exercise 12

Dina and Lila love archery! They shoot arrows at the following target:

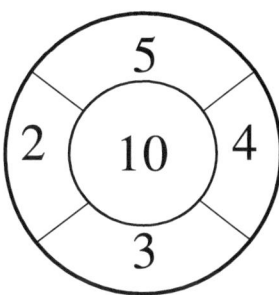

If they hit the target, they gain the number of points written on the portion of the target where the arrow landed. If a player misses the target, 3 points are deducted from her score. After the first three shots, both Dina and Lila have the same score: 12 points. One arrow has missed the target and 5 arrows have hit the target. What is the point difference between each player's highest scoring shot?

Exercise 13

Stephan received a box with 11 pink tennis balls and 16 blue tennis balls. In his studio, he also has a box with 25 older balls, some pink and some blue. What is the largest number of pink balls he could have in total?

Exercise 14

Amira counted her dolls. Since Amira is little, she lost count a few times. She skipped 5, counted 8 twice, skipped 15 and counted 17 three times. When she was finished, she said she had 25 dolls. How many dolls does she really have?

CHAPTER 10. MISCELLANEOUS PRACTICE

Exercise 15

Alfonso, the grocer, has a box with 31 pineapples and mangoes. He puts another 12 pineapples and 7 mangoes in the box. There is now an equal number of pineapples and mangoes in the box. How many mangoes were there in the box before Alfonso added more fruit?

Exercise 16

Dina, Lila, and Amira had to choose their Hallowe'en costumes. They chose from "Selfless Miranda," "Shoeless Cinderella," and "Heartless Fatima." Neither Dina nor Amira were Cinderella, and neither Lila nor Dina were Miranda. Who wore "Heartless Fatima?"

Exercise 17

Amira has lots and lots of stickers with digits on them. One day, she wanted to use some of them to make all the whole numbers from 220 to 240. As she was getting ready to do so, Arbax came in and chewed up the whole stack of "7"s. Amira thought it would be a good idea to put "2"s instead of "7"s throughout. How many different numbers did she end up with?

Exercise 18

Dina places a different digit in each square and in the circle, to satisfy the comparisons. Which digit did she place in the circle?

CHAPTER 10. MISCELLANEOUS PRACTICE

Exercise 19

Lila played around with a drawing tool and made the rectangles in the figure:

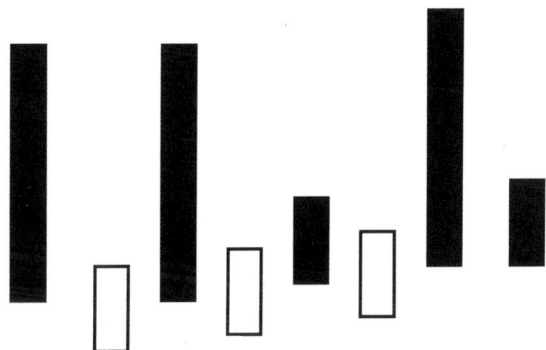

Mark the following statements as either true or false.

(A) If the bar is black, then it is tall.

(B) If the bar is tall, then it is black.

(C) If the bar is short, then it is white.

(D) If the bar is white, then it is short.

CHAPTER 10. MISCELLANEOUS PRACTICE

Exercise 20

Which one of the balloons illustrates the fact that "odd plus odd equals even?"

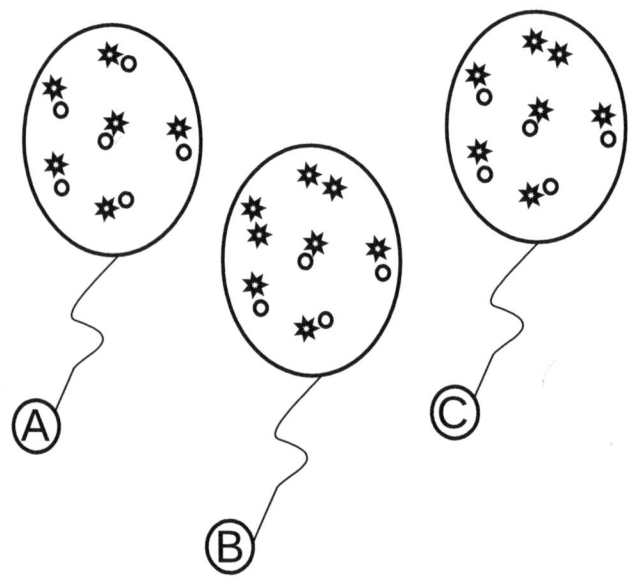

Solutions to Practice One

Exercise 1

Lila has a bracelet made of blue and white magnetic beads:

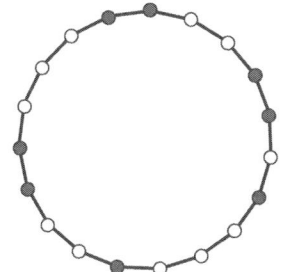

Lila would like to change the bracelet so that all beads of the same color are next to each other. She wants to do this by swapping two beads at a time. What is the smallest number of swaps she must make?

Solution 1

Notice how the two blue beads in the circle fit exactly in the positions occupied by the two white beads indicated by the upper arrow. Similarly, the one lonely blue bead in the triangle fits exactly in the position occupied by the white bead indicated by the lower arrow.

CHAPTER 11. SOLUTIONS TO PRACTICE ONE

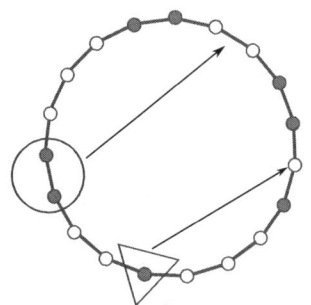

At least three swaps are needed.

Exercise 2

Dina wants to fold the square along one of the grid lines so as to obtain the largest number of matches for the small squares: black faces black and white faces white. What is the largest number of matches she can get by cleverly choosing the line for the fold?

Solution 2

If Dina folds the square along the line shown:

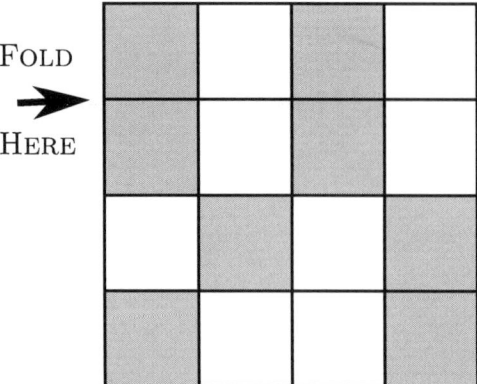

she will obtain 4 matches: 2 black and 2 white.

CHAPTER 11. SOLUTIONS TO PRACTICE ONE

Exercise 3

Lila's dog lives in a rickety old doghouse. It has many holes in the roof! At least how many patches must she use in order to patch all the holes?

Solution 3

Lila notices she can cover two holes with one patch. But, because the number of holes on each slope of the roof is odd, she will have to use an extra patch. For each slope she will use 3 patches for a total of 6 patches.

Exercise 4

Which of the digits in the following picture is the largest?

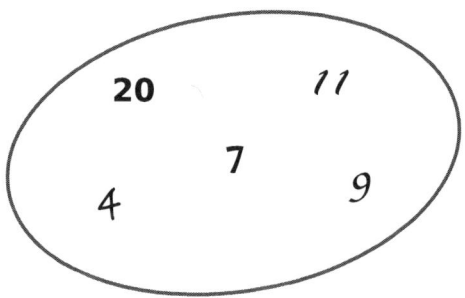

Solution 4

The question is about *digits* not about *numbers*! The largest digit used is 9.

Exercise 5

Out of the pierced chips shown, Dina wants to build the highest tower that will have three holes that match on all chips. How tall will the tower be?

Solution 5

The holes will only match on chips that have 6 and 3 holes. The tallest

CHAPTER 11. SOLUTIONS TO PRACTICE ONE

tower will be formed of 3 chips.

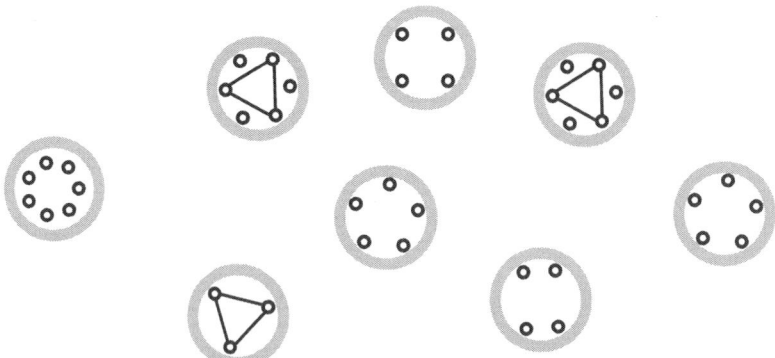

The chip that has 4 holes in a rectangle does have 4 holes that match the holes on the chip with 6 holes but does not match the holes on the chip with 3 holes. We could use the two chips with 6 holes and the chip with the rectangle to build a tower with 4 matching holes that is also 3 chips tall.

Exercise 6

How many even numbers can Lila make by removing some of the digits in the following string? Lila is not allowed to swap any of the remaining digits.

4 7 0 5

Solution 6

Lila can form the even numbers:

4, 0, 40, 70, and 470

for a total of 5 even numbers.

CHAPTER 11. SOLUTIONS TO PRACTICE ONE

Exercise 7

There is a rule for forming the following set of numbers. In one of the numbers, some digits are missing. What could the missing digits be?

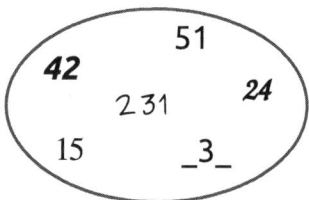

Solution 7

Notice that there are two possible solutions:

(i) The numbers form pairs which have the same digits but in reversed order. The missing number could be 132.

(ii) All the numbers have a sum of digits of 6. The missing number could be one of: 330, 132, and 231.

Exercise 8

Dina is allowed to swap any two adjacent (neighboring) digits in the following string. What is the smallest number of swaps she needs to make in order for all the "3"s to be one beside the other?

Solution 8

Dina can solve this problem in several ways. One possible way is:

3 2 3 1 3 4 3
2 3 3 1 3 4 3
2 3 3 **1 3** 4 3
2 3 3 3 1 **3 4**
2 3 3 3 **3 1** 4

which consists of 4 swaps.

CHAPTER 11. SOLUTIONS TO PRACTICE ONE

Exercise 9

Amira asks her friend to move two candles at the same time and place each of them on a slice neighboring the slice it is on. What is the smallest number of such operations her friend must make in order to move all candles to the same slice?

Solution 9

Two moves are sufficient to place all the candles on the same slice:

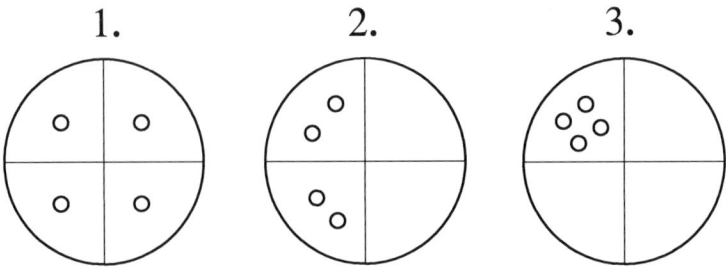

CHAPTER 11. SOLUTIONS TO PRACTICE ONE

Exercise 10

Each of the figures in the diagram follows the same pattern. What number should be placed instead of the question mark?

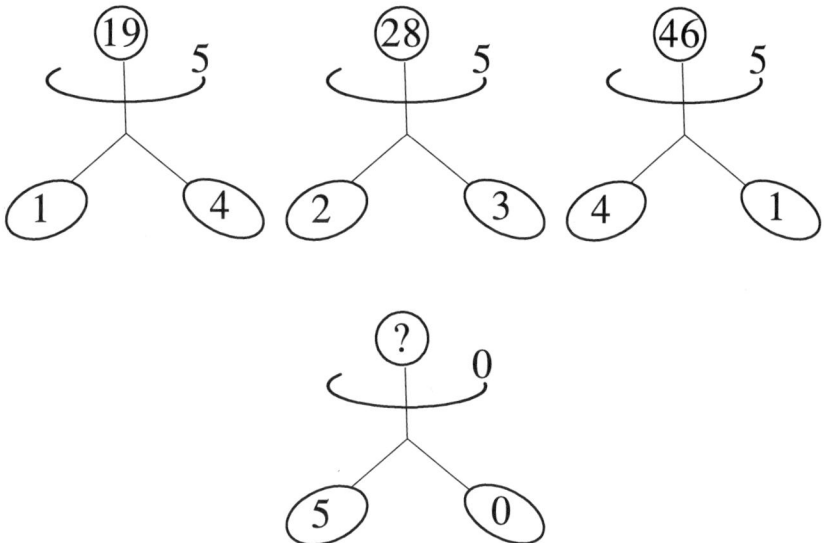

Solution 10

The pattern is to make a two-digit number using the digits placed on the "feet" and add to this number the digit placed in the "hand." The answer is 50.

SOLUTIONS TO PRACTICE TWO

Exercise 1
Dina is making orangeade for 10 of her friends in a large pitcher with a capacity of 8 cups. She has 3 cups of cold water in the pitcher. To this, she adds 7 cups of orange juice. How many cups of orangeade does the pitcher contain now?

Solution 1
Since the pitcher has a capacity of 8 cups, it cannot contain more orangeade than this maximum amount. Dina either spilled some juice or stopped pouring when she noticed the pitcher was full. The number of friends is not relevant.

Exercise 2
Lila has tied her dog Arbax to a small tree using a 6 foot line. When it was time for lunch, Lila placed some bones 10 feet away from Arbax, and went to picnic with Dina nearby. Can Arbax reach the bones?

Solution 2
The bones can be more than 6 feet from Arbax. As long as the bones are within 6 feet of the pole, Arbax will be able to reach them. This is possible in many ways, including this one:

Arbax o———— 6 feet ————|———— 4 feet ————o *Bones*

Pole

CHAPTER 12. SOLUTIONS TO PRACTICE TWO

Exercise 3

Dina and Lila are giving a party for their birthday. They have a bag of party balloons to inflate. Using a small pump, Lila was able to inflate a balloon in 2 minutes. Using a larger pump, Dina was able to inflate a balloon in 1 minute. Which balloon has more air inside?

Solution 3

The two balloons are identical and have approximately the same amount of air inside, regardless of how they were inflated.

Exercise 4

Amira is going down the stairs. Which of her feet is on the lower step: the right one or the left one?

Solution 4

Amira's left foot is on the lower step.

Exercise 5

Dina's phone shows the wrong time. It is 2 hours ahead of local time. If Dina's phone shows it is 11 pm, what is the actual time?

Solution 5

The local time is 9 pm.

Exercise 6

Lila has three boxes and four potatoes. She places the potatoes in the boxes. Then, she makes up a number by counting the potatoes in each box. The leftmost box holds the number of hundreds, the middle box holds the number of tens, and the rightmost box holds the number of units. How many different numbers can she form?

Solution 6

Lila can form the following numbers:
All potatoes are in the rightmost box: 4
All potatoes are in the middle box: 40.
All potatoes are in the leftmost box: 400.

CHAPTER 12. SOLUTIONS TO PRACTICE TWO

Only the leftmost box is empty: 13, 22, 31.
Only the middle box is empty: 103, 301, 202.
Only the rightmost box is empty: 130, 220, 310.
There is at least one potato in each box: 112, 121, 211.

Note to parents: This problem helps students gain hands-on experience with counting the number of ways n objects can be put into $m < n$ boxes.

Exercise 7

Lila is standing in line to buy popcorn at the cinema. There are 5 people in front of her and 3 people behind her. Each time someone buys popcorn and leaves the line, two people join the line. How many people are standing in line by the time Lila gets to order?

Solution 7

14 people: Lila, the 3 people who were behind her at the start of the problem, and the 10 people who lined up behind her while she waited.

Exercise 8

Dina has two cups of hot chocolate milk at 90 degrees Fahrenheit. What is the temperature of each cup?

Solution 8

Each cup has a temperature of 90 degrees Fahrenheit. Temperatures are not added together.

Exercise 9

When the day before yesterday was tomorrow, it was Tuesday. Tomorrow it will be:

(A) Tuesday

(B) Wednesday

(C) Thursday

(D) Friday

(E) Saturday

CHAPTER 12. SOLUTIONS TO PRACTICE TWO

Solution 9

Place *today* on a timeline and model the information in the problem on the timeline.

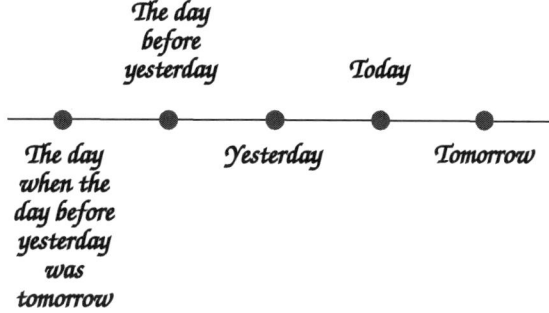

Then, simply mark the days of the week on the timeline in order:

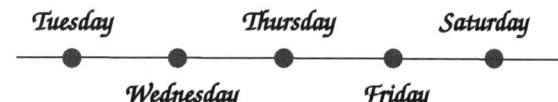

to find out that it will be Saturday.

CHAPTER 12. SOLUTIONS TO PRACTICE TWO

Exercise 10

The purple warbler has a nest full of chicks. Half of her chicks have grey heads, half of her chicks have black tails, and half of her chicks have red bellies. The number of chicks cannot be:

(A) 2

(B) 4

(C) 5

(D) 6

Solution 10

Since the number of chicks must have an integer half, she cannot have a total of 5 chicks. All other choices are possible answers.

Exercise 11

Arbax, the Dalmatian, is 10 steps away from Dina. Dina could be:

(A) anywhere on a 20 step long segment;

(B) anywhere on a 10 step long segment;

(C) at one of the ends of a 20 step long segment;

(D) anywhere on a square with a side 10 steps long;

(E) anywhere on a circle with a radius 10 steps long.

Solution 11

Dina could be anywhere on a circle with a radius 10 steps long.

Exercise 12

Lila and Dina follow the rules:

(i) replace two adjacent pluses with a minus

(ii) replace two adjacent minuses with a plus

(iii) replace two adjacent symbols that are different with a plus

CHAPTER 12. SOLUTIONS TO PRACTICE TWO

to reduce the following string to a single symbol.

$$+-++-$$

Dina works from left to right. Lila works from right to left. Which one of them ends up with a plus: Dina, Lila, or both?

Solution 12

Dina ends up with a plus and Lila ends up with a minus. Dina's moves:
$+++-$
$-+-$
$+-$
$+$
Lila's moves:
$+-++$
$+--$
$++$
$-$

SOLUTIONS TO PRACTICE THREE

Exercise 1

Dina runs 1 mile in 10 minutes. Lila runs half a mile in 6 minutes. Which of them runs faster?

Solution 1

If Dina runs 1 mile in 10 minutes, then she runs half a mile in 5 minutes. Lila runs half a mile in 6 minutes. By comparison, Dina runs faster.

Exercise 2

Arbax, the Dalmatian, barks 3 times every 10 minutes. How many times will Arbax bark in one hour?

Solution 2

Arbax barks 3 times every hour and 3 times every 10, 20, 30, 40, and 50 minutes after the hour. He barks 18 times during one hour.

Exercise 3

Each minute, Dina erases 4 letters from the left while Lila erases 3 letters from the right. After how many minutes will the board be wiped clean?

kavasnexozfboiutymqafhoqetmebwletim

Solution 3

Count all the letters - there are 35 of them. Each minute, Dina and Lila together erase a total of 7 letters. It takes them 5 minutes to wipe the board clean.

CHAPTER 13. SOLUTIONS TO PRACTICE THREE

Exercise 4

It takes Dina 3 minutes to squeeze enough oranges to fill a glass with fresh juice. How many minutes will it take her to fill 5 glasses?

Solution 4

15 minutes

Exercise 5

It takes Lila 3 minutes to hardboil an egg. How long does it take her to hardboil 5 eggs?

Solution 5

3 minutes. Lila can boil all the eggs in the same pot.

Exercise 6

Two kittens can lap up a saucer of milk in 6 minutes. How long will it take four kittens to do the same?

Solution 6

Four kittens will drink the milk faster. They will empty the saucer in 3 minutes.

Exercise 7

One clock is fast and advances 10 minutes more than a regular clock every hour. At 4:00 pm, it is set to show the correct time. When the faster clock is one hour ahead of it, what time will the regular clock show?

Solution 7

It will take 6 hours for the faster clock to be one hour ahead of the regular clock. In six hours' time, it will be 10 pm.

CHAPTER 13. SOLUTIONS TO PRACTICE THREE

Exercise 8

One person wears one pair of pants. How many pairs of pants will two pairs of people wear?

Solution 8

A pair of people is 2 people. Two pairs of people are 4 people. Each person wears one pair of pants. The answer is: 4 pairs of pants.

Exercise 9

It takes Dina three days to knit a scarf and it takes Lila two days to knit a scarf. How many days do the twins need to knit 5 scarves?

Solution 9

Dina knits two scarves in six days. Lila knits three scarves in six days. It takes six days for the girls to knit five scarves.

Exercise 10

Dina has a magical beanstalk that triples in height each day. In six days, the beanstalk reached the height of her house: 18 feet. How many days before that was the beanstalk only 6 feet tall?

Solution 10

Go backwards in time from day to day, dividing the height by 3 each time. Since $18 \div 3 = 6$, the beanstalk must have been 6 feet tall only the day before.

Exercise 11

Over the summer, Lila stays with her cousins while Dina stays with her grandparents. They write each other a letter every day. How many letters will they send in a week?

Solution 11

Dina sends Lila 7 letters, one for each day of the week. Lila also sends 7 letters during that week. In total, they send 14 letters in a week.

CHAPTER 13. SOLUTIONS TO PRACTICE THREE

Exercise 12

A car is 8 times faster than a bicycle. If it takes the car 1 hour to travel a circular path, how many laps can the bicycle complete in 8 hours?

Solution 12

One lap.

Solutions to Practice Four

Exercise 1

The friends of a friend are my friends. This is true for all the people in my circle of friends. If each of us has 8 friends, how many are we?

Solution 1

Since I have 8 friends, there are at least 9 of us. Since this is true for each of us, there are exactly 9 of us.

Exercise 2

Seven frogs are sitting on some waterlilies. Suddenly, a vulture swoops down and snatches one of the frogs. How many frogs are left sitting on the waterlilies?

Solution 2

None. The ones spared by the vulture got scared and scampered off.

Exercise 3

Dina and Lila are holding the baby while their father helps the nurse set up the weighing station. At least how many people are there in the room?

Solution 3

At least five people: Dina, Lila, their father, the nurse, and the baby.

Exercise 4

Dina and Lila traveled by omnibus train from Blois to Tours in 4 hours. If they had used an express train instead, the trip would only have lasted 2 hours. Which train traveled a longer distance?

Solution 4

Both vehicles traveled the same distance: from Blois to Tours.

CHAPTER 14. SOLUTIONS TO PRACTICE FOUR

Exercise 5

Dina baked a large brownie cake and cut it in 12 slices. Lila cut one of the slices in 4. How many slices are there now?

Solution 5

15 slices. Three more slices were added when Lila cut one of the slices in four.

Exercise 6

Lila wants to adopt a cat. At the animal shelter there are 31 cats, of which 15 are tabby and the rest are black. The cats are so cute that Dina also decides to adopt a cat. Dina and Lila walk out, each holding the cat of her choice. What is the largest possible number of black cats that remained at the shelter after Dina and Lila left?

Solution 6

At the shelter there were 16 black cats to start with:

$$31 - 15 = 16$$

Since both Dina and Lila may have chosen tabby cats, the largest possible number of black cats remaining is 16.

Exercise 7

4 race cars covered 4 miles in 1 minute. How many seconds will it take a single race car to cover 4 miles?

Solution 7

Still 1 minute.

CHAPTER 14. SOLUTIONS TO PRACTICE FOUR

Exercise 8

Dina has two friends in grade 2, five friends in grade 4, and four friends in grade 3. At least how many friends does Dina have?

Solution 8

Dina has at least $2 + 5 + 4 = 11$ friends, since none of her friends can be in two different grades simultaneously.

Exercise 9

Four of Lila's friends like Hathi, the elephant. Four of Lila's friends like Rikki, the mongoose. Four of Lila's friends like Felix, the cat. At least how many friends does Lila have?

Solution 9

Lila has at least 4 friends. Each friend might like all the animals.

Exercise 10

Dina has four big magnets and four small magnets. Two magnets are red, one is blue, three are green, and two are yellow. No large magnets are red. No small magnets are blue. An equal number of small and large magnets are painted in the same color. How many large green magnets are there?

Solution 10

The only colors that even numbers of magnets are painted in are red and yellow. Therefore, the group containing the equal number of small and large magnets must have one of these two colors. Since no large magnets are red, yellow is the only option. Hence, one large magnet is yellow and one small magnet is yellow. The blue magnet must be large. There are two large magnets left unaccounted for and they can only be green. There are two large green magnets (and one small green magnet).

CHAPTER 14. SOLUTIONS TO PRACTICE FOUR

Exercise 11

Lila has 28 fish in her 30 liter fish tank. One day, she removes 15 liters of water from the tank. How many fish are there in the tank now?

Solution 11

There will still be 28 fish.

Exercise 12

Dina has more books than Nina. Nina has fewer books than Lila. Edna has more books than Lila. Dina does not have the largest number of books. Who has the smallest number of books?

Solution 12

Make a diagram showing the number of books each girl has in decreasing order. The data is not sufficient to determine this order completely, but may be sufficient to find the answer. Mark the positions that are not completely determined with a question mark:

$$E \quad D \quad L \quad N$$
$$? \quad ?$$

Dina and Lila could be interchanged. Since Edna has more books than Lila and Dina does not have the largest number of books, it follows that Edna must have the largest number of books. Nina, however, has fewer books than any of the others.

Solutions to Miscellaneous Practice

Exercise 1

Dina is in Ms. Left's class at the left end of the schoolyard. Lila is in Ms. Right's class at the right end of the schoolyard. In the middle of the schoolyard, there is a flag. After school, Lila runs towards Dina as Dina walks slowly towards Lila. When they meet, which one of them is closer to the flag?

Solution 1

When they meet, they are the same distance from the flag.

Note to parents: Students must be familiarized with common approximations that are used in word problems. One of them is illustrated by this example: moving objects are generally assumed to be *pointlike*, unless otherwise specified. That is, they do not have a size of their own. Otherwise, we could rightly say that Lila is slightly closer.

Exercise 2

Dina's father is a pilot. His plane can fly at 500 miles per hour. He takes off from Plampaloosa and flies for one hour. After this time, how far away from Plampaloosa can he *not* be?

- **(A)** 0 miles
- **(B)** 200 miles
- **(C)** 300 miles
- **(D)** 500 miles
- **(E)** 600 miles

Solution 2

The plane could not have travelled 600 miles. Notice that the plane can fly upwards, as well as turn back. All distances smaller than or equal to the 500 mile range are possible.

CHAPTER 15. SOLUTIONS TO MISCELLANEOUS PRACTICE

Exercise 3

In Crazy Horse Gulch, there is an unusual weather pattern. It rains for three days, is sunny for three days, rains again for three days, and so on. If it rained yesterday and today, what will the weather be like in 4 days?

(A) rainy for sure

(B) sunny for sure

(C) could be either rainy or sunny

Solution 3

As the diagram shows, there are two possible cases:

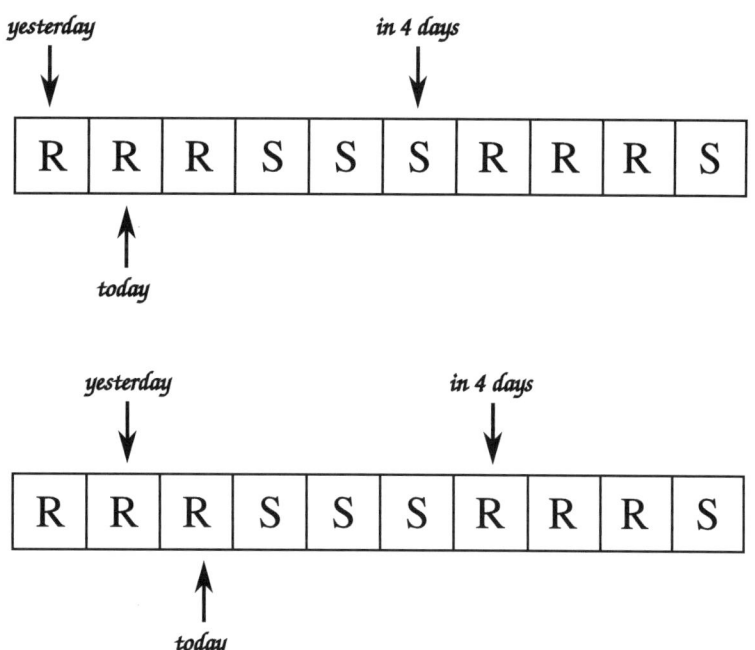

Therefore, it could be either rainy or sunny in four days.

CHAPTER 15. SOLUTIONS TO MISCELLANEOUS PRACTICE

Exercise 4

Lila's father is 30 years older than she is. How many years older than her will her father be when Lila is 30 years old?

Solution 4

The difference in the ages of two people is the same throughout their lives. Lila's father will still be 30 years older than her.

Exercise 5

It takes a magical fig tree 8 years to grow 10 feet tall. How many years does it take 2 magical fig trees to grow 10 feet tall?

Solution 5

Still 8 years. Each tree grows on its own.

Exercise 6

Dina and Lila harvested a lot of figs this summer and decided to dry them out in the sun. If it takes 3 days to dry 100 figs, how many days does it take to dry 300 figs?

Solution 6

3 days. It takes 300 figs just as long to dry as it takes 100 figs.

Exercise 7

Lila and Dina ran towards each other on a 6 mile road. Dina ran twice as fast as Lila. By the time they met, how many miles had Lila run?

Solution 7

Dina ran 4 miles and Lila ran 2 miles until they met ($2 + 4 = 6$).

Exercise 8

An animation begins with a row of 25 squares, all of which are white. Every second, some of the squares turn blue. First, every second square starting from 1 turns blue. Then, every third square starting from 1 turns blue. Then, every fourth square starting from 1 turns blue, and

CHAPTER 15. SOLUTIONS TO MISCELLANEOUS PRACTICE

so on. When will the entire row be blue?

Solution 8

None of the rules makes the second square change color. It will remain white indefinitely.

Exercise 9

A *hexagon* is a closed figure with 6 sides. How many hexagons are hidden in the picture?

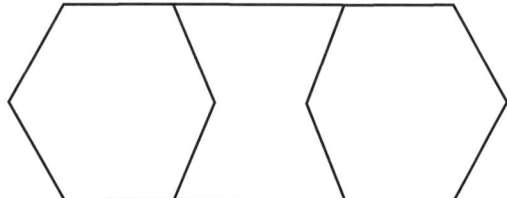

Solution 9

We count 6 hexagons:

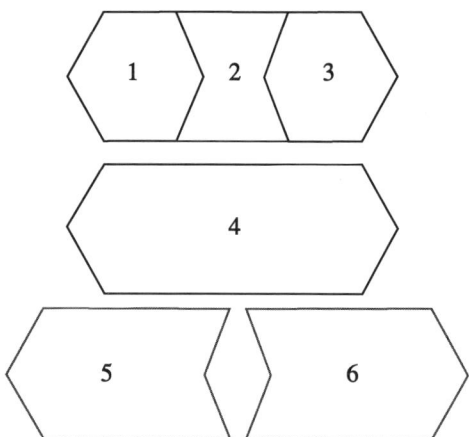

CHAPTER 15. SOLUTIONS TO MISCELLANEOUS PRACTICE

Exercise 10

Dina has lots of these tiles:

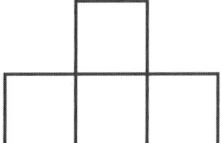

How many of the figures below can she make?

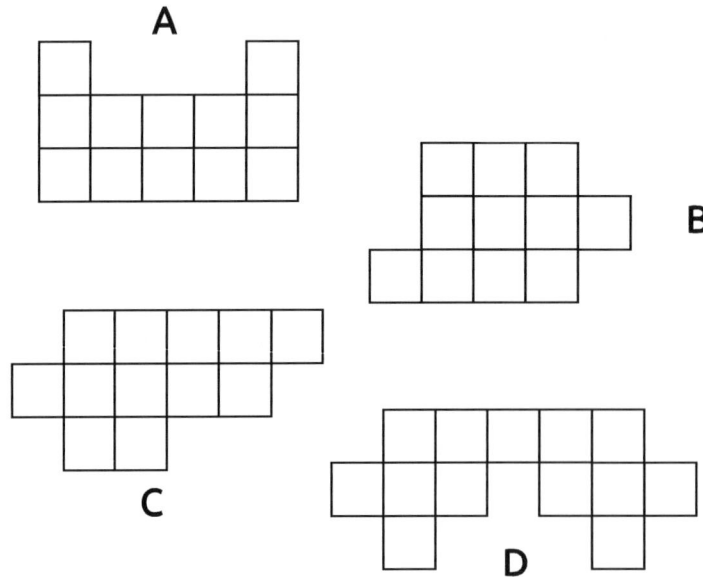

Solution 10

Figures B and D cannot be constructed using Dina's tiles. Only A and C can be built.

CHAPTER 15. SOLUTIONS TO MISCELLANEOUS PRACTICE

Exercise 11

Lila received a robomouse for her birthday. It was designed to walk on a grid either horizontally or vertically. Unfortunately, Lila's robomouse is defective and unable to turn left or go backwards. When Lila places it in the labyrinth at point A, facing in the direction of the arrow, it gets stuck at B. If Lila places it at point B facing in the direction of the arrow, at which of the following points will it get stuck?

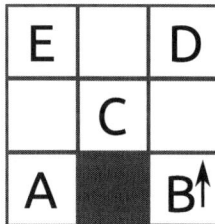

Solution 11

The robomouse will get stuck at D since the only moves possible at D would be to turn left or to go back. The robomouse is defective and cannot turn left or go back.

Exercise 12

Dina and Lila love archery! They shoot arrows at the following target:

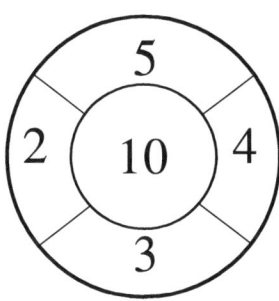

CHAPTER 15. SOLUTIONS TO MISCELLANEOUS PRACTICE

If they hit the target, they gain the number of points written on the portion of the target where the arrow landed. If a player misses the target, 3 points are deducted from her score. After the first three shots, both Dina and Lila have the same score: 12 points. One arrow has missed the target and 5 arrows have hit the target. What is the point difference between each player's highest scoring shot?

Solution 12

It is possible to score 12 points in three shots in one of the three following ways:

$$4 + 5 + 3 = 12$$
$$10 + 5 - 3 = 12$$
$$5 + 5 + 2 = 12$$

The highest scores are 10 and 5. The difference between them is 5.

Exercise 13

Stephan received a box with 11 pink tennis balls and 16 blue tennis balls. In his studio, he also has a box with 25 older balls, some pink and some blue. What is the largest number of pink balls he could have in total?

Solution 13

Since we know for sure that in the old box not all the balls are pink, there are at most 24 pink balls in it. Add the 11 new pink balls to find the largest possible number of pink balls: $24 + 11 = 35$.

Exercise 14

Amira counted her dolls. Since Amira is little, she lost count a few times. She skipped 5, counted 8 twice, skipped 15 and counted 17 three times. When she was finished, she said she had 25 dolls. How many dolls does she really have?

Solution 14

Skipping a number is like counting a doll she does not have. For instance, if she counts "second doll, fourth doll," she actually has only 3

CHAPTER 15. SOLUTIONS TO MISCELLANEOUS PRACTICE

dolls at this point. Since she skips two numbers, we must decrease her total by 2: $25 - 2 = 23$.

Counting a number twice means that Amira has a doll but fails to account for it and gives it the same number as the previous doll. Therefore, we must add 1 to the total: $23 + 1 = 24$.

Counting a number three times means that Amira has two dolls she fails to account for and gives the same number as the previous doll. Therefore, we must add 2 to the total: $24 + 2 = 26$.

Amira has 26 dolls.

Exercise 15

Alfonso, the grocer, has a box with 31 pineapples and mangoes. He puts another 12 pineapples and 7 mangoes in the box. There is now an equal number of pineapples and mangoes in the box. How many mangoes were there in the box before Alfonso added more fruit?

Solution 15

Add all the fruit together to find the total: $31 + 12 + 7 = 50$. Of these, half (25) are pineapples and half are mangoes. If Alfonso had to add 7 mangoes to reach the count of 25, then there were $25 - 7 = 18$ mangoes in the box before he added more fruit.

Exercise 16

Dina, Lila, and Amira had to choose their Hallowe'en costumes. They chose from "Selfless Miranda," "Shoeless Cinderella," and "Heartless Fatima." Neither Dina nor Amira were Cinderella, and neither Lila nor Dina were Miranda. Who wore "Heartless Fatima?"

Solution 16

If Dina and Amira are not Cinderella, then Lila must be Cinderella. If Lila and Dina are not Miranda, then Amira must be Miranda. This leaves Dina as Fatima.

CHAPTER 15. SOLUTIONS TO MISCELLANEOUS PRACTICE

Exercise 17

Amira has lots and lots of stickers with digits on them. One day, she wanted to use some of them to make all the whole numbers from 220 to 240. As she was getting ready to do so, Arbax came in and chewed up the whole stack of "7"s. Amira thought it would be a good idea to put "2"s instead of "7"s throughout. How many different numbers did she end up with?

Solution 17

Without Arbax's intervention, Amira would have created $240 - 220 + 1 = 21$ numbers. Because of the change, 227 is turned into 222 and 237 is turned into 232. 222 and 232 are numbers in the set that Amira wants to create. Therefore, they will appear twice. The number of different numbers is $21 - 2 = 19$.

Exercise 18

Dina places a different digit in each square and in the circle, to satisfy the comparisons. Which digit did she place in the circle?

Solution 18

The total number of squares and circles is 10. This means that all 10 digits must be used. Though there are many ways of placing digits in the squares, only zero can be placed in the circle. We show two different solutions in the figures. From these, one can see how more solutions can be built.

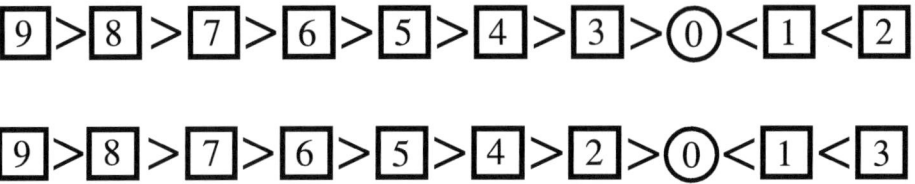

Regardless of how we arrange the digits in the squares, the digit in the circle has to be smaller than all other digits.

CHAPTER 15. SOLUTIONS TO MISCELLANEOUS PRACTICE

Exercise 19

Lila played around with a drawing tool and made the rectangles in the figure:

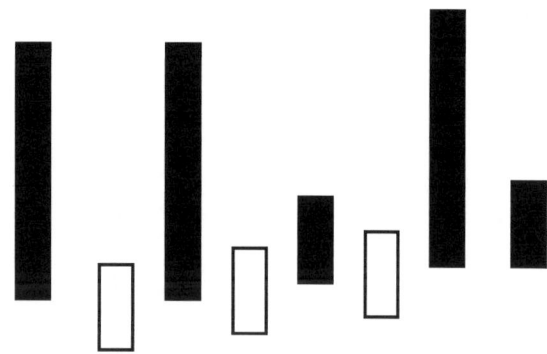

Mark the following statements as either true or false.

(A) If the bar is black, then it is tall.

(B) If the bar is tall, then it is black.

(C) If the bar is short, then it is white.

(D) If the bar is white, then it is short.

Solution 19

(A) False. Some black bars are short.

(B) True. All tall bars are black.

(C) False. Some short bars are black.

(D) True. All white bars are short.

CHAPTER 15. SOLUTIONS TO MISCELLANEOUS PRACTICE

Exercise 20

Which one of the balloons illustrates the fact that "odd plus odd equals even?"

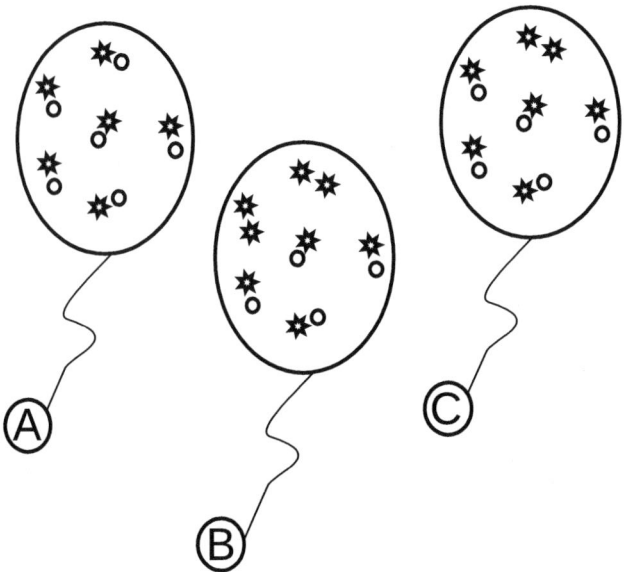

Solution 20

Since it is possible to pair the objects in each balloon, the total number of objects in each balloon is even. The three balloons illustrate the two different cases of addition that can produce an even sum:

(A) $6 + 6 = 12$, "Even plus even equals even."

(B) $8 + 4 = 12$, "Even plus even equals even."

(C) $5 + 7 = 12$, "Odd plus odd equals even."

The answer is (C).

Competitive Mathematics Series for Gifted Students

Practice Counting (ages 7 to 9)
Practice Logic and Observation (ages 7 to 9)
Practice Arithmetic (ages 7 to 9)
Practice Operations (ages 7 to 9)

Practice Word Problems (ages 9 to 11)
Practice Combinatorics (ages 9 to 11)
Practice Arithmetic (ages 9 to 11)
Practice Operations (ages 9 to 11)

Practice Word Problems (ages 11 to 13)
Practice Combinatorics (ages 11 to 13)
Practice Arithmetic and Number Theory (ages 11 to 13)
Practice Algebra and Operations (ages 11 to 13)
Practice Geometry (ages 11 to 13)

Practice Word Problems (ages 12 to 15)
Practice Algebra and Operations (ages 12 to 15)
Practice Geometry (ages 12 to 15)
Practice Number Theory (ages 12 to 15)
Practice Combinatorics and Probability (ages 12 to 15)

This is a series of practice books. With the exception of a few reminders, there are no theoretical explanations. For lessons, please see the resources indicated below:

Find a set of free lessons in competitive mathematics at www.mathinee.com. Addressing grades 5 through 11, the *Math Essentials* on www.mathinee.com present important concepts in a clear and concise manner and provide tips on their application. The site also hosts over 400 original problems with full solutions for various levels. Selectors enable the user to sort essentials and problems by test or contest targeted as well as by topic and by the earliest grade level they can be used for.

Online problem solving seminars are available at www.goodsofthemind.com. If you found this booklet useful, you will enjoy the live problem solving seminars.

For supplementary assessment material, look up our problem books in test format. The "Practice Tests in Math Kangaroo Style" are fun to use and have a well organized workflow.

Made in the USA
San Bernardino, CA
01 March 2015